MW00390492

No-Nonsense General Class License Study Guide

for tests given given between July 2015 and June 2019

Dan Romanchik, KB6NU

ISBN: 0692433104
ISBN-13: 978-0692433102

CONTENTS

UPGRADING TO GENERAL CLASS

As you probably know by now, there are three amateur radio license classes: Technician, General, and Amateur Extra. The Technician Class license is the license that most beginners get, and many amateur radio operators never go any further. While there are plenty of fun things that you can do with a Technician Class license, upgrading to General will allow you to do many more things. Most notably, when you pass the General Class license test, legally referred to as Element 3, you can operate phone and digital modes in parts of all the HF (shortwave) bands.

Like the Technician Class test, the General Class test is a 35-question, multiple-choice examination. You must score at least 75% on the test, meaning that you must answer 26 of the 35 questions correctly.

The test is more difficult from a technical point of view than the Technician Class license, but you don't need an engineering degree to pass the test. With a little bit of study, using this study guide, you can feel confident about passing the test.

To obtain a General Class license, you must pass the Technician Class test in addition to the General Class test. If you don't already have a Technician Class license, you might want to download the No-Nonsense, Technician Class License Study Guide from http://www.kb6nu.com/study-guides/.

Where do I take the test?

Amateur radio license examinations are given by Volunteer Examiners, or VEs. VEs are licensed radio amateurs who have been trained to administer amateur radio tests. To find out when the VEs in your area will be giving the test go to the American Radio Relay League's (ARRL). On the Exam Session Search page (http://www.arrl.org/arrl/vec/examsearch.phtml), you will be able to search for test sessions that are close to you. If you do not have access to the Internet, you can phone the ARRL at 860-594-0200.

Can I really learn how to be an amateur radio operator using this simple study guide?

Yes and no. This study guide will help you get your license, but getting your license is only the beginning. There is still much to learn, and to get the most out of your General Class license, you will have to continually learn new things. This study guide will teach you the answers to the test questions, but not give you a deep understanding of electronics, radio, or the rules and regulations.

One way that you can explore the topics in this study guide further is to do Google searches as you read a particular section. For example, as you read the section on Yagi antennas, search for "yagi antenna" to get more detailed explanations and to find images of Yagi antennas. You may even find YouTube videos that show you how to build a Yagi antenna.

Of course, after you get your license, you'll be able to use it to good effect. I hope that by helping you get your license you'll be encouraged to become an active radio amateur and get on the air, participate in public service and emergency communications, join an amateur radio club, and experiment with radios, antennas, and circuits. These are the activities that will really help you learn about radio in depth, and in the end, help you be confident in your abilities as an amateur radio operator.

How do I use this manual?

Simply read through the manual, search the Internet as noted above, and take some online practice tests. You will find the answers to questions in bold. Question designators, such as "(G5A07)" appear at the end of sentences. This is so you can refer to actual question in the question pool, if you would like to. You can take practice tests online at QRZ.Com, AA9PW.Com, and several other websites.

Good luck and have fun!

I hope that you find this study guide useful and that you'll upgrade to General. The General Class license will allow you to do more things, meaning that you'll learn new things.

If you have any comments, questions, compliments or complaints, I want to hear from you. E-mail me at cwgeek@kb6nu.com. My goal is to continually refine this study guide and to continually make it better.

73!

Dan Romanchik KB6NU
cwgeek@kb6nu.com, Twitter: @kb6nu

DAN ROMANCHIK, KB6NU

ELECTRICAL PRINCIPLES

Reactance; inductance; capacitance; impedance; impedance matching

In direct-current (DC) circuits, resistance opposes the flow of current. In alternating current (AC) circuits, both capacitors and inductors oppose the flow of current. Reactance is **opposition to the flow of alternating current caused by capacitance or inductance**. (G5A02) We use the letter X to stand for reactance. **Ohm** is the unit used to measure reactance. (G5A09)

Reactance causes opposition to the flow of alternating current in an inductor. (G5A03) The reactance caused by a capacitor or inductor depends on the frequency of the AC source. You calculate the reactance caused by an inductor with this equation:

$$X_L = 2\pi f L$$

where X_L is the inductive reactance, f is the frequency of the AC source, and L is the inductance in henries. **As the frequency of the applied AC increases, the reactance of an inductor increases.** (G5A05)

Reactance causes opposition to the flow of alternating current in a capacitor. (G5A04) You calculate the reactance caused by a capacitor with this equation:

$$X_C = 1/(2\pi fC)$$

where X_C is the capacitive reactance, f is the frequency of the AC source, and C is the capacitance in farads. **As the frequency of the applied AC increases, the reactance of a capacitor decreases**. (G5A06)

Resistors also oppose the flow of current in an AC circuit. When an AC circuit contains both resistance and reactance, we call the combination of the two impedance. Impedance is **the opposition to the flow of current in an AC circuit**. (G5A01) **Ohm** is the unit used to measure impedance. (G5A10)

When setting up an amateur radio station, it is important to know the input and output impedances of devices and circuits that you will connect together. When these two impedances are equal, they are said to "match" one another. For example, when the output impedance of a transmitter is 50 ohms and the input impedance of an antenna is 50 ohms, they match one another.

If the output impedance of a circuit or device (often called the "source") does not match the input impedance of the circuit or device that you connect to it (often called the "load"), the source will not deliver the maximum amount of power to the load. Impedance matching is, therefore, important **so the source can deliver maximum power to the load**. (G5A08) When the impedance of an electrical load is equal to the internal impedance of the power source, **the source can deliver maximum power to the load**. (G5A07)

Impedance matching is so important that engineers have devised several different types of circuits and devices to match impedances. **All of these choices are correct** when talking about devices that can be used for impedance matching at radio frequencies (G5A13):

- A transformer
- A Pi-network
- A length of transmission line

One method of impedance matching between two AC circuits is to **insert an LC network between the two circuits**. (G5A11) LC circuits consist of inductors and capacitors and are often used between two circuits operating at radio frequencies, or RF.

You can also use a matching transformer between two RF circuits. One reason to use an impedance matching transformer is **to maximize the transfer of power**. (G5A12)

The Decibel; current and voltage dividers; electrical power calculations; sine wave root-mean-square (RMS) values; PEP calculations

Kirchoff's Current Law states that the sum of currents entering a circuit node must equal the sum of the currents leaving the node. Consequently, it's not hard to see that the total current entering a parallel circuit **equals the sum of the currents through each branch.** (G5B02)

The RMS value, or root mean square value, of an AC signal is the voltage that causes the same power dissipation as a DC voltage of the same value. (G5B07) For an AC signal with a sine-wave shape, the RMS value is .707 times the peak value. **12 volts** is the RMS voltage of a sine wave with a value of 17 volts peak. (G5B09)

Power is equal to the RMS voltage times the current, or

$$P \text{ (watts)} = V_{RMS} \times I$$

Using Ohm's Law, we can show that:

$$P = V^2_{RMS} / R$$

$$P = I^2_{RMS} \times R$$

Using these formulas, you can see that **200 watts** of electrical power are used if 400 VDC is supplied to an 800-ohm load (G5B03):

$$P = V^2_{RMS} / R = 400^2/800 = 160,000 / 800 = 200 \text{ watts}$$

2.4 watts of electrical power are used by a 12-VDC light bulb that draws 0.2 amperes (G5B04):

$$P \text{ (watts)} = V_{RMS} \times I = 12 \times 0.2 = 2.4 \text{ watts}$$

Approximately 61 milliwatts are being dissipated when a current of 7.0 milliamperes flows through 1.25 kilohms (G5B05):

$$P = I^2_{RMS} \times R = 0.007^2 \times 1,250 \approx .061 \text{ W} \approx 61 \text{ milliwatts}$$

These formulas can also be used to calculate RF power and RF voltages and currents. **245 volts** would be the voltage across a 50-ohm dummy load dissipating 1200 watts (G5B12):

$$V^2_{RMS} = P / R = 50 \times 1,200 = 60,000, \sqrt{60,000} \approx 245 \text{ volts}$$

A term sometimes used to describe the power output of a phone signal is peak envelope power (PEP). This is the maximum instantaneous power achieved when transmitting a phone signal. For an unmodulated carrier, though, the peak envelope power is equal to the average power. Therefore, **1060 watts** is the output PEP of an unmodulated carrier if an average reading wattmeter connected to the transmitter output indicates 1060 watts. (G5B13) The ratio of peak envelope power to average power for an unmodulated carrier is **1.00**. (G5B11) When transmitting a single sideband signal, though, the amplitude will vary with time, and the average power may be considerably less.

The output PEP from a transmitter is **100 watts** if an oscilloscope measures 200 volts peak-to-peak across a 50-ohm dummy load connected to the transmitter output. (G5B06) **625 watts** is the output PEP from a transmitter if an oscilloscope measures 500 volts peak-to-peak across a 50-ohm resistor connected to the transmitter output. (G5B14)

Often, we're not concerned with the actual power, but with the ratio of power input to power output. For example, if an amplifier has a gain of 10, we know that if we input a 1 W signal, we'll get 10 W out. Quite often, you'll see this ratio specified in decibels, or dB.

The formula for calculating power ratios in dB is:

$$A(dB) = 10 \times \log_{10} (P_2 / P_1)$$

Using this formula, you can see that a two-times increase or decrease in power results in a change of **3 dB** (G5B01):

$$A(dB) = 10 \times \log_{10} (P_2/P_1) = 10 \times \log_{10} (2/1) = 10 \times .301 \approx 3\ dB$$

By rearranging the terms of the equation, you would calculate that a power loss of **20.5 %** would result from a transmission line loss of 1 dB (G5B10):

$$P_2/P_1 = 10^{(A(dB)/10)} = 10^{-0.1} = .795$$

Note that since this is a power loss, the value we use in the equation is -1 dB, and the power loss is 20.5% because P_2/P_1 is equal to 79.5%.

Resistors, capacitors, and inductors in series and parallel; transformers

Connecting components in series and in parallel will affect their effective values. For example, if you connect resistors in series, the effective resistance is the sum of the individual resistances. **A resistor in series** should be added to an existing resistor in a circuit to increase circuit resistance. (G5C03)

Connecting resistors in parallel will decrease the circuit resistance. The equation that you use to calculate the total resistance of resistors in parallel is:

$$R_{total} = 1/(1/R_1 + 1/R_2 + 1/R_3...)$$

When the resistors are all equal, simply divide the value of one of the resistors by the number of resistors in parallel. For example, the total resistance of three 100-ohm resistors in parallel is **33.3 ohms** (G5C04):

$$R_{total} = 100 / 3 = 33.3 \text{ ohms}$$

5.9 ohms is the total resistance of a 10 ohm, a 20 ohm, and a 50 ohm resistor in parallel (G5C15):

$$R_{total} = 1/(1/10 + 1/20 + 1/50) = 1/0.17 \approx 5.9 \text{ ohms}$$

150 ohms is the value of each resistor which, when three of them are connected in parallel, produce 50 ohms of resistance, and the same three resistors in series produce 450 ohms. (G5C05)

Inductors work the same way as resistors. **An inductor in series** should be added to an inductor in a circuit to increase the circuit inductance. (G5C14) The inductance of a 20 millihenry inductor in series with a 50 millihenry inductor is **70 millihenrys** (G5C11), but the inductance of three 10 millihenry inductors connected in parallel is **3.3 millihenrys**. (G5C10)

Capacitors, however, are quite the opposite. **A capacitor in**

parallel should be added to a capacitor in a circuit to increase the circuit capacitance, (G5C13) while connecting capacitors in series will decrease circuit capacitance. The capacitance of a 20 microfarad capacitor in series with a 50 microfarad capacitor is **14.3 microfarads**. (G5C12) The capacitance of three 100 microfarad capacitors connected in series **33.3 microfarads**. (G5C09)

When working with capacitors, it's important to take note of how the capacitance is specified. The capacitance is sometimes specified in microfarads (μF), sometimes in nanofarads (nF), and sometimes in picofarads (pF). A microfarad is 1,000 nanofarads, and a nanofarad is 1,000 picofarads. So, for example, the value in nanofarads of a 22,000 pF capacitor is **22 nF**. (G5C17) The value in microfarads of a 4700 nanofarad (nF) capacitor is **4.7 μF**. (G5C18)

Where it's really important to keep this in mind is when you're working with some capacitors whose capacitance is specified in nanofarads and others who are specifed in picofarads. For example, the equivalent capacitance of two 5.0 nanofarad capacitors and one 750 picofarad capacitor connected in parallel is **10.750 nanofarads**. (G5C08) 750 pF is equal to 0.75 nF, and because the capacitors are connected in parallel, the equivalent capacitance is equal to the sum of all the capacitances.

Inductors exhibit a behavior called mutual inductance. Mutual inductance occurs when a current   flowing through one inductor induces a current in a nearby inductor. We use this behavior to create components called transformers.

The simplest transformer has two windings: a primary winding and a secondary winding. When an AC voltage source is connected across its primary winding, **mutual inductance** causes a voltage to appear across the secondary winding of a transformer. (G5C01)

The voltage across the secondary winding will be equal to the voltage across the primary times the ratio of the number of turns in the secondary to the number of turns in the primary.   When the number of turns in the secondary winding is greater than the number of turns in the primary winding, the voltage across the secondary winding will be great than the voltage across the primary winding, and the transformer is called a step-up transformer.

When the number of turns in the secondary winding is less than the number of turns in the primary winding, the voltage across the secondary winding will be less than the voltage across the primary winding, and the transformer is called a step-down transformer. For example, the voltage across a 500-turn secondary winding of a transformer is **26.7 volts** if the 2250-turn primary is connected to 120 VAC. (G5C06)

By reversing a transformer's windings, that is connecting an input voltage to a transformer's secondary winding and connecting the primary windings to the output circuit, you make a step-up transformer act like a step-down transformer and vice versa. For example, if you reverse the primary and secondary windings of a 4:1 voltage step down transformer, **the secondary voltage becomes 4 times the primary voltage**. (G5C02) In effect, it becomes a step-up transformer.

Doing this is not necessarily a good idea, however. Current in the primary winding of a step-up transformer is higher than the current in the secondary, and **to accommodate the higher current of the primary**, the conductor of the primary winding of many voltage step-up transformers is larger in diameter than the conductor of the secondary winding. (G5C16) If you use a step-down transformer as a step-up transformer by connecting the input voltage to the secondary winding, the wire in the winding may not be able to handle the higher current and burn out.

Transformers are also used to transform impedances. The impedance ratio is also related to the turns ratio, but the transformation is equal to the square of the turns ratio. The turns ratio of a transformer used to match an audio amplifier having a 600-ohm output impedance to a speaker having a 4-ohm impedance is **12.2 to 1**. (G5C07)

DAN ROMANCHIK, KB6NU

CIRCUIT COMPONENTS

Resistors; capacitors; Inductors; rectifiers; solid state diodes and transistors; vacuum tubes; batteries

There are a number of practical considerations you must make when using electronic components in circuits. For example, the resistance of a resistor **will change depending on the resistor's temperature coefficient** if the temperature is increased. (G6A16)

One type of resistor that's commonly used in electronics is the wire-wound resistor. Its main advantage is that its value can be set very precisely. You probably don't want to use wire-wound resistors in RF circuits, though. A reason not to use wire-wound resistors in an RF circuit is that **the resistor's inductance could make circuit performance unpredictable**. (G6A17)

It's also important to choose capacitors wisely. For example, the primary advantage of ceramic capacitors is **comparatively low cost**. (G6A14) A disadvantage, though, is that they may have a high temperature coefficient.

High capacitance for given volume is an advantage of an electrolytic capacitor. (G6A15) For this reason, electrolytic capacitors are often used in power-supply circuits to filter the rectified AC.

Electrolytic capacitors are polarized, meaning that they have both positive and negative leads. That being the case, you must be careful

15

not to connect them so that the voltage on the positive lead is negative with respect to the voltage on the negative lead. **All of these choices are correct** when considering why the polarity of applied voltages is important for polarized capacitors (G6A13):

- Incorrect polarity can cause the capacitor to short-circuit
- Reverse voltages can destroy the dielectric layer of an electrolytic capacitor
- The capacitor could overheat and explode

A popular choice for inductors is the ferrite core inductor. **All of these choices are correct** when talking about the advantages of using a ferrite core with a toroidal inductor (G6A18):

- Large values of inductance may be obtained
- The magnetic properties of the core may be optimized for a specific range of frequencies
- Most of the magnetic field is contained in the core

Placing two inductors closely together may cause unwanted mutual inductance. It is important to minimize the mutual inductance between two inductors to reduce unwanted coupling between circuits, which could cause the circuits to malfunction. The winding axes of two solenoid inductors should be placed **at right angles to each other** to minimize their mutual inductance. (G6A10)

The diode is one of the most versatile components. They are used to rectify AC voltages, regulate DC voltages, switch RF voltages, and demodulate radio signals. One of the most important diode specifications is the junction threshold voltage. This voltage will be different for different types of diodes. The approximate junction threshold voltage of a germanium diode is **0.3 volts** (G6A03). The approximate junction threshold voltage of a silicon diode is **0.7 volts**. (G6A05)

A type of diode often used in RF circuits is the Schottky diode. **Lower capacitance** is an advantage of using a Schottky diode in an RF switching circuit as compared to a standard silicon diode. (G6A06) The lower capacitance enables it to switch faster than other types of diodes.

There are two main types of transistors, bipolar transistors and field effect transistors (FETs). **A field effect transistor** is the solid state device most like a vacuum tube in its general operating characteristics. (G6A11) One type of FET is the Metal Oxide Semiconductor FET, or MOSFET. In a MOSFET, **the gate is separated from the channel with a thin insulating layer**. (G6A09)

Transistors are often used in amateur radio circuits as amplifiers and switches. When used in an amplifier or as a power supply pass transistor, the cases of some large power transistors must be insulated from ground **to avoid shorting the collector or drain voltage to ground**. (G6A08) When used as a switch in a logic circuit, the stable operating points for a bipolar transistor are **its saturation and cut-off regions**. (G6B07)

Some amateur radio equipment, most notably linear amplifiers, still use vacuum tubes. The simplest vacuum tube is the triode, which has three elements: the cathode, the control grid, and the plate. The **control grid** is the element of a triode vacuum tube used to regulate the flow of electrons between cathode and plate. (G6A10)

A pentode is a tube with five elements, one of which is called the screen grid. The primary purpose of a screen grid in a vacuum tube is **to reduce grid-to-plate capacitance**. (G6B12)

Batteries are often used to power amateur radio equipment, so it's important to know their characteristics and how to use them. Some batteries can be used only once, while others are rechargeable. Nickel-cadmium batteries, sometimes called "Ni-Cads," are one type of rechargeable battery. **High discharge current** is an advantage of the low internal resistance of Nickel Cadmium batteries. (G6A02)

Carbon-zinc batteries, on the other hand, are not rechargeable. It is **never** acceptable to recharge a carbon-zinc primary cell. (G6A04)

For applications that require long battery life, 12 volt lead-acid batteries, often car batteries or deep cycle marine batteries, are used. When using these batteries, it's important to not discharge them all the way. **10.5 volts** is the minimum allowable discharge voltage for maximum life of a standard 12 volt lead-acid battery. (G6A01)

Analog and digital integrated circuits (IC's); micropro-cessors; memory; I/O devices; microwave IC's (MMIC's); display devices

Most amateur radio devices today contain integrated circuits (ICs). On a single piece of silicon, manufacturers can put thousands of transistors, implementing very complex circuit functions.

ICs may contain analog circuits, digital circuits, or a combination of both. A simple example is the **linear voltage regulator**, which is an analog integrated circuit (G6B01). **Analog** is also the term that describes an integrated circuit operational amplifier. (G6B06)

A more sophisticated analog IC is the MMIC. The term MMIC means **Monolithlic Microwave Integrated Circuit.** (G6B02)

Digital ICs are often described by the technology used to create the transistors. One type, or "family," of digital IC uses complementary, metal-oxide semiconductor, or CMOS, transistors. CMOS has several advantages over other IC technologies, such as transistor-transistor logic (TTL). **Low power consumption** is an advantage of CMOS integrated circuits compared to TTL integrated circuits. (G6B03)

Memories are an important type of digital IC. One type of memory is the read-only memory, or ROM, which is often used to store programs or data that never changes. The term ROM means **Read Only Memory.** (G6B04) ROM is characterized as "non-volatile," meaning **the stored information is maintained even if power is removed.** (G6B05)

ROMs are often used to store programs that run microprocessors. A microprocessor is **a computer on a single integrated circuit**. (G6B11) Modern amateur radio transceivers usually have one or more microprocessors to control their operation.

Modern transceivers use light-emitting diodes (LEDs) or liquid crystal displays (LCDs) to display operating status, such as frequency. An LED is **forward biased** when emitting light. (G6B08) Older radios often used incandescent bulbs instead of LEDs. **Lower power consumption** is one advantage of an LED when compared to an incandescent indicator. (G6B07)

LCDs do not emit light. Therefore, one of the characteristics of a liquid crystal display is that **it requires ambient or back lighting**. (G6B09)

Most modern transceivers can now be controlled by a computer. A common way to connect a computer to a transceiver is with a USB interface. A **computer and transceiver** are two devices in an amateur radio station that might be connected using a USB interface. (G6B10)

To connect all of the devices we typically have in an amateur radio station, we use many different types of connecters. For example, a **DE-9** connector would be a good choice for a serial data port. (G6B12) An **RCA Phono** connector is commonly used for audio signals in Amateur Radio stations. (G6B14)

Many connectors were designed specifically to connect RF signals. A **PL-259** connector is commonly used for RF service at frequencies up to 150 MHz. (G6B13) The type-N connector is **a moisture-resistant RF connector useful to 10 GHz**. (G6B16) The SMA connector is **a small threaded connector suitable for signals up to several GHz**. (G6B18)

To help users make sure they make the right connections, some connectors are keyed. The main reason to use keyed connectors instead of non-keyed types is **reduced chance of incorrect mating**. (G6B15) DIN connectors are one type of keyed connector. The general description of a DIN type connector is a **family of multiple circuit connectors suitable for audio and control signals**. (G6B17)

DAN ROMANCHIK, KB6NU

PRACTICAL CIRCUITS

Power supplies and schematic symbols

Power supplies are devices that convert AC power to the DC voltages needed to power amateur radio equipment. There are two main types of power supplies available: linear power supplies and switching power supplies.

Linear supplies use a transformer to transform the voltage up or down, a rectifier to convert the AC voltage to a DC voltage, and capacitors and inductors to smooth the output voltage. The rectifier in a linear supply may be a half-wave rectifier, a full-wave rectifier, or a bridge rectifier.

180 degrees is the portion of the AC cycle that is converted to DC by a half-wave rectifier (G7A05). The peak-inverse-voltage across the rectifiers in a half-wave power supply is **two times the normal peak output voltage of the power supply**. (G7A04)

360 degrees is the portion of the AC cycle that is converted to DC by a full-wave rectifier (G7A06). The peak-inverse-voltage across the rectifiers in a full-wave bridge power supply is **equal to the normal peak output voltage of the power supply**. (G7A03). **A series of DC pulses at twice the frequency of the AC input** is the output waveform of an unfiltered full-wave rectifier connected to a resistive load (G7A07).

The output of a rectifier connects to a filter made up of capacitors and inductors. **Capacitors and inductors** are used in a power-supply filter network (G7A02). A component often found across the output of a power supply is a power-supply bleeder resistor. A power supply bleeder resistor is a safety feature in that **it discharges the filter capacitors**. (G7A01). The value of this resistor is normally very high value so that very little current flow throug it during normal operation.

Switching, or switched-mode power supplies are now being sold by many vendors. One advantage of a switched-mode power supply as compared to a linear power supply is that **high frequency operation allows the use of smaller components** (G7A08). One disadvantage is that the circuits are much more complex than linear power supply circuits.

When designing or troubleshooting radios, amateur radio operators use schematic diagrams to describe circuits. Various symbols represent the different types of components. A typical schematic is shown in Figure G7-1.

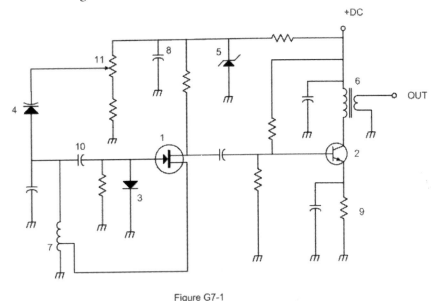

Figure G7-1

Symbol 1 in figure G7-1 represents a field effect transistor. (G7A09)

Symbol 5 in figure G7-1 represents a Zener diode. (G7A10)

Symbol 2 in figure G7-1 represents an NPN junction transistor. (G7A11)

Symbol 6 in Figure G7-1 represents a multiple-winding transformer. (G7A12)

Symbol 7 in Figure G7-1 represents a tapped inductor. (G7A13)

Digital circuits; amplifiers and oscillators

Digital circuits are circuits whose output are one of two voltages—either "on" or "off," "high" or "low," "one" or "zero." Digital circuits use the binary system to represent numbers because each of the digits in a binary number is either a 1 or a 0. An advantage of using the binary system when processing digital signals is that **binary "ones" and "zeros" are easy to represent by an "on" or "off" state**. (G7B02)

We use digital circuits to implement logic functions and there are many integrated circuits that implement specific logic functions, such as AND and NOR. For a two input AND gate, the **output is high only when both inputs are high**. (G7B03) For a two input NOR gate, the **output is low when either or both inputs are high**. (G7B04)

Integrated circuits that provide more complex logic functions, such as counters and shift registers, are also available. A 3-bit binary counter has **8** states. (G7B05) The figure below shows a three-bit counter made with 3 D flip-flops. A D flip-flop is a circuit whose output, Q, changes when it receives a clock pulse. After the clock pulse, Q is equal to D. The other output, Q[bar] is the inverse of Q.

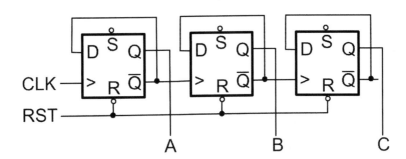

A three-bit counter has eight output states.

24

A shift register is **a clocked array of circuits that passes data in steps along the array.** (G7B06)

Complex digital circuitry can often be replaced by a **microcontroller.** (G7B01) Microcontrollers can be programmed in much the same way that you program a personal computer. The advantage to this approach is that instead of rewiring a circuit, you simply modify the microcontroller's program. Microcontrollers, such as the PIC line of microcontrollers, are available for less than a dollar each.

Oscillators and amplifiers

An oscillator is a circuit that generates an AC output signal. The basic components of virtually all sine wave oscillators are **a filter and an amplifier operating in a feedback loop.** (G7B07)

An "LC" oscillator uses an inductor and a capacitor connected so that they form what's called a tank circuit to provide feedback. **The inductance and capacitance in the tank circuit** determines the frequency of an LC oscillator. (G7B09)

There are many different types of amplifiers. **An amplifier in which the output preserves the input waveform** is called a linear amplifier. (G7B14) Linear amplifiers are usually Class A amplifiers. **Low distortion** is a characteristic of a Class A amplifier. (G7B10) They are, therefore, most appropriate for amplifying phone signals.

The Class C amplifier is not linear. A Class C power stage is appropriate for amplifying a **CW** modulated signal. (G7B11) **Class C** amplifiers have the highest efficiency. (G7B12) To determine the efficiency of an RF power amplifier, **divide the RF output power by the DC input power.** (G7B08)

High-power amplifiers are often prone to self-oscillation due to stray capacitive feedback. To prevent this from happening, you induce some feedback that is out of phase with the stray capacitive feedback to neutralize it. The reason for neutralizing the final amplifier stage of a transmitter is **to eliminate self-oscillations.** (G7B13)

Receivers and transmitters, filters, oscillators

Filters are very important circuits in amateur radio equipment. As the name implies, these circuits are used to clarify or process radio signals. For example, one type of filter—a low-pass filter—passes all signals whose frequencies are below a certain frequency, called the "cutoff frequency."

One application of a low-pass filter is to block the VHF and UHF harmonics produced by an amateur transceiver from reaching the antenna. To do this, you would connect the input of the filter to the output of your transceiver and the output of the filter to your antenna system. When used in this way, the impedance of a low-pass filter should be **about the same** as the impedance of the transmission line into which it is inserted. (G7C06)

Filters are also used in amateur radio transmitters. A **filter** is used to process signals from the balanced modulator and send them to the mixer in a single-sideband phone transmitter. (G7C01) A **balanced modulator** is the circuit used to combine signals from the carrier oscillator and speech amplifier and send the result to the filter in a typical single-sideband phone transmitter. (G7C02)

These days, many transceivers use digital circuits, instead of analog circuits, to filter RF signals. This technique is called Digital Signal Processing (DSP). Digital Signal Processor filtering is accomplished **by converting the signal from analog to digital and using digital processing**. (G7C10) What's happening is that a specialized computer chip, called a Digital Signal Processor, is running software that performs many millions of calculations on the digital representation of the signal.

All of the these choices are correct when talking about what is needed for a Digital Signal Processor IF filter (G7C09):
- An analog to digital converter
- A digital to analog converter
- A digital processor chip

The superheterodyne receiver is the most popular type of amateur radio receiver. Superheterodyne receivers convert the received signal to an intermediate frequency (IF) and then process that IF signal. A **mixer** is the circuit used to process signals from the RF amplifier and local oscillator and send the result to the IF filter in a superheterodyne receiver. (G7C03) A **product detector** is the circuit used to combine signals from the IF amplifier and BFO and send the result to the AF amplifier in some single-sideband receivers. (G7C04) The simplest combination of stages that implement a superheterodyne receiver is **HF oscillator, mixer, detector**. (G7C07)

FM receivers have different types of circuits than the superheterodyne receivers designed for AM, CW, and SSB reception. A **discriminator** circuit is used in many FM receivers to convert signals coming from the IF amplifier to audio. (G7C08)

Most modern transceivers use digital circuits called synthesizers to control the receive and transmit frequencies. An advantage of a transceiver controlled by a direct digital synthesizer (DDS) is that it provides **variable frequency with the stability of a crystal oscillator**. (G7C05)

Digital techniques have proven to be so effective at generating and receiving radio signals, that some transceivers now implement most functions using digital signal processors. We even have a special term for these types of radios. The term "software defined radio" (SDR) means **a radio in which most major signal processing functions are performed by software**. (G7C11)

SIGNALS AND EMISSIONS

Carriers and modulation: AM; FM, single and double sideband, modulation envelope, overmodulation

Modulation is the process by which we convey information over a radio link, and amplitude modulation (AM) is perhaps the most basic type of modulation. When we amplitude modulate a signal, what we're doing is varying the amplitude of what we call the RF carrier signal in proportion to the amplitude of an audio signal.

Amplitude modulation is, therefore, the type of modulation that varies the instantaneous power level of the RF signal. (G8A05) The modulation envelope of an AM signal is **the waveform created by connecting the peak values of the modulated signal**. (G8A11) The figure below shows an amplitude modulated signal and the modulation envelope.

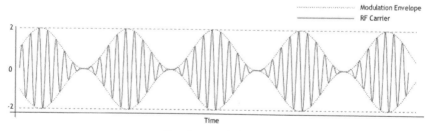

An amplitude-modulated signal.

Single sideband, or SSB, is a type of amplitude modulation. A conventional AM signal has three components, the carrier and two sidebands. Since the carrier signal carries no information and the information in both sidebands is the same, someone figured out that by eliminating the carrier and one of the sidebands you could still transfer the same amount of information.

Not only that, SSB voice communications has a couple of advantages of AM voice communications. One advantage of carrier suppression in a single-sideband phone transmission is that **the available transmitter power can be used more effectively**. (G8A06) The reason for this is that you're putting all of the output power into the single sideband and not wasting it for the carrier and extraneous sideband.

Another advantage is that the phone emission that uses the narrowest frequency bandwidth is **single sideband**. (G8A07) An AM signal requires a bandwidth of 6 kHz or more; a single sideband signal takes up a bandwidth of only 3 kHz.

You must be careful when setting the audio level used to modulate a phone signal, no matter if it's AM or SSB. If you set the level too high, the signal will be over-modulated, and this may cause "flat-topping." Flat-topping of a single-sideband phone transmission is **signal distortion caused by excessive drive**. (G8A10) Another effect of over-modulation is **excessive bandwidth** (G8A08).

To set the appropriate audio level, you adjust the microphone gain control while watching the radio's automatic level control, or ALC, meter. The **transmit audio or microphone gain** control is typically adjusted for proper ALC setting on an amateur single sideband transceiver. (G8A09)

Another type of modulation that we use in amateur radio is frequency modulation, or FM. Most VHF and UHF repeaters use frequency modulation. **Frequency modulation** is the name of the process which changes the frequency of an RF wave to convey information. (G8A03) In an FM signal, the carrier frequency changes proportionally to the instantaneous amplitude of the modulating signal.

Another way to produce a frequency-modulated signal is to use phase modulation. **Phase modulation** is the name of the process that changes the phase angle of an RF wave to convey information. (G8A02) **Phase modulation** is produced by a reactance modulator connected to an RF power amplifier. (G8A04)

Some modes, such as radio teletype (RTTY) use a modulation method called frequency-shift keying, or FSK. An FSK signal is generated **by changing an oscillator's frequency directly with a digital control signal**. (G8A01)

Frequency mixing, multiplication, bandwidths of various modes, deviation; duty cycle

One of the most important circuits found in amateur radio equipment is the mixer. A mixer takes two input signals and outputs the sum and difference of the two input signals. **Heterodyning** is another term for the mixing of two RF signals. (G8B03)

The **mixer** is the receiver stage that combines a 14.250 MHz input signal with a 13.795 MHz oscillator signal to produce a 455 kHz intermediate frequency (IF) signal. (G8B01) If a receiver mixes a 13.800 MHz VFO with a 14.255 MHz received signal to produce a 455 kHz intermediate frequency (IF) signal, a 13.345 MHz signal will produce an **image response** in the receiver. (G8B02)

FM transmitters use multipliers to produce a VHF signal. The **multiplier** is the name of the stage in a VHF FM transmitter that generates a harmonic of a lower frequency signal to reach the desired operating frequency. (G8B04) When modulating the oscillator, you must use a proportionally smaller deviation if you plan to multiply the oscillator's output to the 2m band. For example, **416.7 Hz** is the frequency deviation for a 12.21-MHz reactance-modulated oscillator in a 5-kHz deviation, 146.52-MHz FM-phone transmitter. (G8B07)

FM phone is often used on the VHF and UHF bands, but not below 29.5 MHz because it requires a fair amount of bandwidth. **16 kHz** is the total bandwidth of an FM-phone transmission having a 5 kHz deviation and a 3 kHz modulating frequency. (G8B06)

"Digital modes," such as PACTOR3, are getting more popular on the HF bands. The bandwidth of a PACTOR3 signal at maximum data rate is approximately **2300 Hz**. (G8B05). In general, digital modes have much higher duty cycles than the traditional HF modes, such as CW or SSB. When operating digital modes, it is important to know the duty cycle of the data mode you are using when transmitting because **some modes have high duty cycles which could exceed the transmitter's average power rating**. (G8B08)

It's also important to know the symbol rate. Digital modes that transfer data at a high rate need more bandwidth than modes that send data at a slower rate. The relationship between transmitted symbol rate and bandwidth is that **higher symbol rates require higher bandwidth**. (G8B10)

Noise can be a problem when operating digital modes because noise can cause errors, and many of these digital modes do not have error correction. One way to minimize the effects of noise is to use filters to filter out the noise and to adjust your receiver's bandwidth so that it only passes the signal that you're interested in. It is good to match receiver bandwidth to the bandwidth of the operating mode because **it results in the best signal to noise ratio**. (G8B09)

Digital emission modes

Digital modes have become very popular in amateur radio. To use the digital modes, you connect a computer your radio, and the computer is responsible for encoding and decoding the digital signal.

PSK 31 is perhaps the most popular digital mode on the HF bands. The number 31 in "PSK31" represents **the approximate transmitted symbol rate**. (G8C09) That's pretty slow, but it is about as fast as a person can type.

Varicode is the type of code is used for sending characters in a PSK31 signal. (G8C12) It is named Varicode because **the number** of data bits sent in a single PSK31 character **varies**. (G8C02) **Upper case letters use longer Varicode signals and thus slow down transmission**. (G8C08)

Perhaps the second-most popular digital mode is radio teletype, or RTTY. In the past, we used clunky, mechanical teletype machines to send and receive RTTY, but now most amateurs use a computer to operate this mode.

RTTY uses the Baudot code, which is described as **a 5-bit code with additional start and stop bits**. (G8C04) RTTY uses frequency-shift keying to send the ones and zeroes. One frequency is interpreted as a one, or "mark." The other frequency is interpreted as a zero or "space." **Mark and Space** identifies the two separate frequencies of a Frequency Shift Keyed (FSK) signal. (G8C11)

Packet radio is a form of digital switching technology used by several different digital modes, including APRS and PACTOR. In packet radio, data transmitted from one station to another are bundled up in packets. The **header** of a data packet contains the routing and handling information. (G8C03)

PACTOR is a particularly robust mode that uses packet radio technology to extend Internet functionality to remote users via amateur radio. The reason amateurs use PACTOR is that it detects errors and has a mechanism for correcting them. In the PACTOR protocol, an NAK response to a transmitted packet means **the**

receiver is requesting the packet be retransmitted. (G8C05) This occurs when the receiving station has detected a data error. When using PACTOR, **the connection is dropped** when the stations fail to exchange information due to excessive transmission attempts. (G8C06)

Another method for controlling errors when sending and receiving data packets is called Automatic Repeat Request, or ARQ. A station receiving an ARQ data mode packet containing errors **requests the packet be retransmitted**. (G8C07)

Some digital modes, such as FreeDV use what's called forward error correction, which allows the receiving station to not only detect errors in a data packet, but also to correct them. Forward error correction (FEC) allow the receiver to correct errors in received data packets **by transmitting redundant information with the data**. (G8C10) The obvious disadvantage of this approach is that it is slower than other data modes because redundant data needs to be sent.

Two digital modes that are becoming more popular are JT9 and JT95. **JT9 and JT65** are the digital modes designed to operate at extremely low signal strength on the HF bands. (G8C01) While these modes work very well, they do require that both stations are synchronized to one another.

ANTENNAS AND FEED LINES

Antenna feed lines: characteristic impedance, attenuation, SWR calculation, measurement and effects, matching networks

Feedlines are the cables used to connect antennas to receivers and transmitters. The most important characteristic of a feedline is its characteristic impedance. **The distance between the centers of the conductors and the radius of the conductors** determine the characteristic impedance of a parallel conductor antenna feed line. (G9A01)

50 and 75 ohms are the typical characteristic impedances of coaxial cables used for antenna feed lines at amateur stations. (G9A02) The reason we use cables with these impedances is that they closely match the impedance of commonly used amateur radio antennas, such as quarter-wave verticals (35 ohms) and half-wave dipoles (72 ohms). **300 ohms** is the characteristic impedance of flat ribbon TV type twinlead. (G9A03)

A difference between feed-line impedance and antenna feed-point impedance is the reason for the occurrence of reflected power at the point where a feed line connects to an antenna. (G9A04) A measure of this mismatch is the voltage standing-wave ratio, or simply SWR. The SWR is equal to the ratio of the impedances.

A standing wave ratio of **1:1** will result from the connection of a

50-ohm feed line to a non-reactive load having a 50-ohm impedance. (G9A11) This is the best possible case. When the SWR is 1:1, we say that the feedline is "matched" to the load. To prevent standing waves on an antenna feed line, **the antenna feed-point impedance must be matched to the characteristic impedance of the feed line**. (G9A07)

When the two impedances are not matched, an SWR greater than 1:1 will result, and the SWR will be equal to the ratio between the two impedances. For example, a **4:1** standing wave ratio will result from the connection of a 50-ohm feed line to a non-reactive load having a 200-ohm impedance. (G9A09) A standing wave ratio of **5:1** will result from the connection of a 50-ohm feed line to a non-reactive load having a 10-ohm impedance. (G9A10)

If you feed a vertical antenna that has a non-reactive 25-ohm feed-point impedance with 50-ohm coaxial cable, the SWR will be **2:1**. (G9A12) If you feed an antenna that has a purely resistive 300-ohm feed-point impedance with 50-ohm coaxial cable, the SWR will be **6:1**. (G9A13)

In order not to damage your transmitter, it's important that the impedance its output "sees" is 50 ohms. To accomplish this, we often use devices called antenna tuners, and when adjusted properly, they transform the impedance at the end of the feedline to 50 ohms. That makes the transmitter happy, but the SWR on the feedline is unchanged. If the SWR on an antenna feed line is 5 to 1, and a matching network at the transmitter end of the feed line is adjusted to 1 to 1 SWR, the resulting SWR on the feed line is still **5 to 1**. (G9A08)

When the SWR on a coaxial cable feedline is greater than 1:1, it will attenuate the signal because there will be some interaction between high standing wave ratio (SWR) and transmission line loss. **If a transmission line is lossy, high SWR will increase the loss.** (G9A14) To transfer the greatest amount of power from the transmitter to the receiver, the SWR on the feedline should be 1:1.

Even when perfectly matched, a coaxial cable will attenuate the signal somewhat, depending on the frequency of the signal. Coaxial cable **attenuation increases** as the frequency of the signal it is carrying increases. (G9A05) RF feed line losses are usually expressed in **decibels per 100 ft**. (G9A06)

This transmission line loss will have an effect on the SWR measured at the input to the line. **The higher the transmission line loss, the more the SWR will read artificially low.** (G9A15)

Basic antennas

There are many different types of antennas, including:

- random-wire antennas,
- dipole antennas, and
- vertical antennas, including ground plane antennas.

As the name implies, random-wire antennas are a random-length. To match the antenna to the transmitter, you'll need an antenna tuner, which is normally located in the shack. Because of this, there may be high RF levels in the shack when you are transmitting. One disadvantage, therefore, of a directly fed random-wire antenna is that **you may experience RF burns when touching metal objects in your station**. (G9B01)

The half-wavelength dipole antenna is perhaps the most common amateur radio antenna because it is simple to build and operate. In practice, the half-wave dipole antenna is a bit shorter than a half wavelength. This is due to the effect of the ground and nearby objects on the antenna.

The formula most often used by radio amateurs to calculate the length of a dipole antenna is:

Length (feet) = 468 / f (MHz)

The approximate length for a 1/2-wave dipole antenna cut for 3.550 MHz is **131 feet**. (G9B11)

L = 468 / 3.55 ≈ 131 feet

32 feet is the approximate length for a 1/2-wave dipole antenna cut for 14.250 MHz. (G9B10)

L = 468 / 14.250 ≈ 32 feet

When the feedpoint is at the center of the antenna, the impedance is approximately 72 ohms, making it a good match for 75-ohm coax and 50-ohm coax. The feed-point impedance of a 1/2 wave dipole **steadily increases** as the feed-point location is moved from the center toward the ends. (G9B08)

Dipole antennas are usually mounted horizontally. An advantage of a horizontally polarized HF antenna, as compared to a vertically polarized antenna, is **lower ground reflection losses**. (G9B09)

Ideally, a dipole antenna should be mounted a half-wavelength up off the ground. The low angle azimuthal radiation pattern, or the radiation pattern in the plane of the antenna, of an ideal half-wavelength dipole antenna installed 1/2 wavelength high and parallel to the Earth **is a figure-eight at right angles to the antenna**. (G9B04) **If the antenna is less than 1/2 wavelength high, the azimuthal pattern is almost omnidirectional**. (G9B05)

Antenna height also affects the feed-point impedance. As the antenna is lowered from 1/4 wave above ground, the feed-point impedance of a 1/2 wave dipole antenna **steadily decreases**. (G9B07)

A vertical antenna is a quarter-wavelength long and operates against ground or a set of radials. The approximate length for a 1/4-wave vertical antenna cut for 28.5 MHz is **8 feet**. (G9B12)

When mounted above ground and used with radials, the vertical antenna is called a ground plane antenna. The radial wires of a ground-mounted vertical antenna system should be placed **on the surface or buried a few inches below the ground**. (G9B06) This is to reduce ground losses.

The natural feed-point impedance of a quarter-wave vertical is 35 ohms, but the feed-point impedance of a ground-plane antenna **increases** when its radials are changed from horizontal to downward-sloping. (G9B03). A common way, therefore, to adjust the feed-point impedance of a quarter wave ground-plane antenna to be approximately 50 ohms is to **slope the radials downward**. (G9B02)

Directional antennas

To make their signals more effective, some amateurs use directional antennas. Directional antennas, such as Yagis and quads, direct most of the power output in a particular direction, making the signal seem more powerful. They are also more sensitive to receiving signals from a particular direction.

The "gain" of a directional antenna is the relative increase in power radiated in the direction in which the antenna is pointing. The gain is usually specified in decibels, or dB. Look at this specification very carefully, because the gain may be specified in relation to either an isotropic antenna (dBi) or in relation to a dipole (dBd). When referring to antenna gain, **dBi refers to an isotropic antenna, dBd refers to a dipole antenna.** (G9C20)

dBi is strictly a theoretical figure, as the isotropic antenna is strictly a theoretical construction. dBd is a more realistic specification. When stated in dBi, the gain of an antenna will always be higher than if it is stated in dBd. **dBi gain figures are 2.15 dB higher than dBd gain figures**. (G9C19)

A characteristic related to the antenna gain is the "front-to-back ratio." The "front-to-back ratio" of a Yagi antenna is **the power radiated in the major radiation lobe compared to the power radiated in exactly the opposite direction**. (G9C07) The "major lobe" or "main lobe" of a directive antenna is **the direction of maximum radiated field strength from the antenna**. (G9C08)

Yagis are perhaps the most common type of directional antenna. A Yagi antenna consists of a driven element, a reflector, and one or more directors. The reflector and directors are called parasitic elements. The approximate length of the driven element of a Yagi antenna is **1/2 wavelength**. (G9C02) **The reflector is normally the longest parasitic element** of a three-element, single-band Yagi antenna. (G9C04) In a three-element, single-band Yagi antenna, **the director is normally the shortest parasitic element**. (G9C03)

By changing the physical characteristics of the elements and the spacing between the elements, you can change the characteristics of the antenna. For example, **larger diameter elements** increase the bandwidth of a Yagi antenna. (G9C01) The **gain increases** when you increase boom length and add directors to a Yagi antenna. (G9C05)

All of these choices are correct when talking about Yagi antenna design variables that could be adjusted to optimize forward gain, front-to-back ratio, or SWR bandwidth (G9C10):

- The physical length of the boom
- The number of elements on the boom
- The spacing of each element along the boom

While a Yagi antenna is a great antenna, you can improve the performance of this antenna by stacking one on top of another. The gain of two 3-element horizontally polarized Yagi antennas spaced vertically 1/2 wavelength apart typically is **approximately 3 dB higher** than the gain of a single 3-element Yagi. (G9C09) The advantage of vertical stacking of horizontally polarized Yagi antennas is that it **narrows the main lobe in elevation**. (G9D05)

Although the driven element of a Yagi antenna is similar to a dipole, the other elements cause the feedpoint impedance to be significantly lower than 72 ohms. The purpose of a gamma match used with Yagi antennas is **to match the relatively low feed-point impedance to 50 ohms**. (G9C11) An advantage of using a gamma match for impedance matching of a Yagi antenna to 50-ohm coax feed line is that **it does not require that the elements be insulated from the boom**. (G9C12)

You can also make directional antennas using loop antenna elements. The elements of a quad antenna are square loops. Each side of a quad antenna driven element is approximately **1/4 wavelength**. (G9C13) Each side of a quad antenna reflector element is **slightly more than 1/4 wavelength**. (G9C15) Assuming one of the elements is used as a reflector, **the reflector element must be approximately 5 percent longer than the driven element**, for the antenna to operate as a beam antenna. (G9C06)

The forward gain of a two-element quad antenna is **about the** same as the forward gain of a three-element Yagi antenna. (G9C14) **The polarization of the radiated signal changes from horizontal to vertical** when the feed point of a quad antenna is changed from the center of either horizontal wire to the center of either vertical wire. (G9C18)

The elements of a delta loop beam are triangular. Each leg of a symmetrical delta-loop antenna is approximately **1/3 wavelength**. (G9C17) The gain of a two-element delta-loop beam is **about the same** as the gain of a two-element quad antenna. (G9C16)

Specialized antennas

Another type of directional antenna is the log-periodic antenna. It is called this because for a log periodic antenna, the **length and spacing of the elements increases logarithmically from one end of the boom to the other**. (G9D07) The gain of a log periodic antenna is less than that of a Yagi, but an advantage of a log periodic antenna is **wide bandwidth**. (G9D06)

The term "NVIS" means **Near Vertical Incidence Sky wave** when related to antennas. (G9D01) An NVIS antenna is typically installed **between 1/10 and 1/4 wavelength** above ground. (G9D03) An advantage of an NVIS antenna is **high vertical angle radiation for working stations within a radius of a few hundred kilometers**. (G9D02)

A Beverage antenna is **a very long and low directional receiving antenna**. (G9D10) An application for a Beverage antenna is **directional receiving for low HF bands**. (G9D09) A Beverage antenna is not used for transmitting because **it has high losses compared to other antennas**. (G9D08)

Many antennas are designed for a single band, but in many cases, putting up an antenna for each band you want to operate is impractical. So, many amateurs put up antennas that will work on more than one band. These are called multiband antennas. A disadvantage of multiband antennas is that **they have poor harmonic rejection**. (G9D11)

One type of multiband antenna is the trap vertical. Antenna traps block RF energy in a certain frequency band. This makes the antenna look shorter than it really is at that frequency. The primary purpose of antenna traps is **to permit multiband operation**. (G9D04)

RADIO WAVE PROPAGATION

Sunspots and solar radiation, ionospheric disturbances, propagation forecasting, and indices

Amateur radio communications are subject to the whims of nature. Many different phenomena affect the propagation of signals, and it behooves you to know a little something about the phenomena. Doing so will make you a more effective amateur radio communicator.

The phenomenon that most affects amateur radio communications on the HF bands is the sunspot cycle. The typical sunspot cycle is approximately **11 years** long. (G3A11) During a cycle, the number of sunspots varies from none to a high of between 100 and 200.

The sunspot number (SSN) is very significant when it comes to predicting HF propagation. **Higher sunspot numbers generally indicate a greater probability of good propagation at higher frequencies.** (G3A01) The effect that high sunspot numbers have on radio communications is that **long-distance communication in the upper HF and lower VHF range is enhanced.** (G3A09)

Because counting sunspots is a relatively subjective measure of solar activity, scientists have come up with a more objective measurement, called solar flux. The solar-flux index is **a measure of solar radiation at 10.7 cm.** (G3A05) As the SSN varies from 0 to around 200, the solar flux varies from around 60 to 300.

At any point in the solar cycle, the 20 meter band usually supports worldwide propagation during daylight hours. (G3A07) **15 meters, 12 meters, and 10 meters** are the amateur radio HF bands that are least reliable for long distance communications during periods of low solar activity. (G3A04)

The sunspot cycle is a long-term phenomenon. There are other phenomena that affect radio wave propagation in the short term. For example, **the Sun's rotation on its axis** causes HF propagation conditions to vary periodically in a 28-day cycle. (G3A10)

One phenomenon that can have a drastic effect on propagation is a Sudden Ionic Disturbance (SID). During an SID, the sun emits a great deal of ultraviolet and X-ray radiation. **8 minutes** is approximately how long it takes for the increased ultraviolet and X-ray radiation from solar flares to affect radio-wave propagation on the Earth. (G3A03) The effect a Sudden Ionospheric Disturbance has on the daytime ionospheric propagation of HF radio waves is that **it disrupts signals on lower frequencies more than those on higher frequencies**. (G3A02)

Also, **HF communications are disturbed** by the charged particles that reach the Earth from solar coronal holes. (G3A14) It takes **20 to 40 hours** for charged particles from coronal mass ejections to affect radio-wave propagation on the Earth. (G3A15)

Geomagnetic activity, such as a geomagnetic storm, can also affect radio propagation. A geomagnetic storm is **a temporary disturbance in the Earth's magnetosphere**. (G3A06) One of the effects a geomagnetic storm can have on radio-wave propagation is **degraded high-latitude HF propagation**. (G3A08) **Auroras that can reflect VHF signals** are a possible benefit to radio communications resulting from periods of high geomagnetic activity. (G3A16)

There are two indices that give an indication of the stability of the Earth's magnetic field. The K-index indicates **the short term stability of the Earth's magnetic field**. (G3A12) The A-index indicates **the long term stability of the Earth's geomagnetic field**. (G3A13)

Maximum Usable Frequency, Lowest Usable Frequency, propagation

The two most important parameters for predicting the propagation between two locations are the MUF and LUF. MUF stands for **the Maximum Usable Frequency for communications between two points**. (G3B08) LUF stands for **the Lowest Usable Frequency for communications between two points**. (G3B07)

When they are sent into the ionosphere, radio waves with frequencies below the Maximum Usable Frequency (MUF) and above the Lowest Usable Frequency (LUF) **are bent back to the Earth**. (G3B05) When they are sent into the ionosphere, radio waves with frequencies below the Lowest Usable Frequency (LUF) **are completely absorbed by the ionosphere**. (G3B06) **No HF radio frequency will support ordinary skywave communications over the path** when the Lowest Usable Frequency (LUF) exceeds the Maximum Usable Frequency (MUF). (G3B11)

All of these choices are correct when talking about factors that affect the Maximum Usable Frequency (MUF) (G3B12):
- Path distance and location
- Time of day and season
- Solar radiation and ionospheric disturbances

When selecting a frequency for lowest attenuation when transmitting on HF, **select a frequency just below the MUF**. (G3B03) A reliable way to determine if the Maximum Usable Frequency (MUF) is high enough to support skip propagation between your station and a distant location on frequencies between 14 and 30 MHz is to **listen for signals from an international beacon in the frequency range you plan to use**. (G3B04)

While signals most often take the shortest path from point to point, sometimes the best path for radio propagation is in the opposite direction, also called the "long path." **A well-defined echo might be heard** if a sky-wave signal arrives at your receiver by both short path and long path propagation. (G3B01)

The 6m band is a favorite of many amateur radio operators, even though it infrequently supports long-distance, skywave propagation. A good indicator of the possibility of sky-wave propagation on the 6 meter band is that there is **short skip sky-wave propagation on the 10 meter band**. (G3B02)

Ionospheric layers, critical angle and frequency, HF scatter, Near Vertical Incidence Sky-wave

The ionosphere is what makes long-distance radio communications possible on the shortwave bands. The ionosphere is made up of three layers of charged particles, labelled D, E, and F. **Where the Sun is overhead,** ionospheric layers reach their maximum height. (G3C02)

During the day, there are two F layers, F1 and F2. The F2 region is mainly responsible for the longest distance radio wave propagation **because it is the highest ionospheric region**. (G3C03) **2,500 miles** is the approximate maximum distance along the Earth's surface that is normally covered in one hop using the F2 region. (G3B09) At night, F1 and F2 combine into a single F layer.

While the F2 layer is the layer that normally reflects HF radio signals, the E layer can also reflect signals under some conditions. **1,200 miles** is the approximate maximum distance along the Earth's surface that is normally covered in one hop using the E region. (G3B10)

The ionospheric layer closest to the surface of the Earth is **the D layer**. (G3C01) **The D layer** is the ionospheric layer that is the most absorbent of long skip signals during daylight hours on frequencies below 10 MHz. (G3C12) Long distance communication on the 40, 60, 80 and 160 meter bands is more difficult during the day because **the D layer absorbs signals at these frequencies during daylight hours**. (G3C05)

One factor that affects how well the ionosphere will reflect a signal is the angle at which the signal impinges upon it. If the angle is too high, it will pass right through the ionosphere and not be reflected back to earth. **The highest takeoff angle that will return a radio wave to the Earth under specific ionospheric conditions** is called the critical angle. (G3C04)

Antennas used for DXing should have low takeoff angles. One thing that affects the takeoff angle of an antenna is its height above ground. **A horizontal dipole placed between 1/8 and 1/4**

wavelength above the ground will be most effective for skip communications on 40 meters during the day. (G3C11)

One interesting propagation phenomenon is scatter propagation. **Scatter** propagation allows a signal to be detected at a distance too far for ground wave propagation but too near for normal sky-wave propagation. (G3C09) An indication that signals heard on the HF bands are being received via scatter propagation is that **the signal is heard on a frequency above the Maximum Usable Frequency**. (G3C10) HF scatter signals in the skip zone are usually weak because **only a small part of the signal energy is scattered into the skip zone**. (G3C08)

A characteristic of HF scatter signals is that **they have a wavering sound**. (G3C06) HF scatter signals often sound distorted because **energy is scattered into the skip zone through several different radio wave paths**. (G3C07)

Another interesting phenomenon is Near Vertical Incidence Skywave propagation. Near Vertical Incidence Sky-wave (NVIS) propagation is **short distance HF propagation using high elevation angles**. (G3C13) Basically what happens is that the antenna sends the signal at an angle close to 90 degrees, and if conditions are right, the ionosphere reflects that signal back to the earth at a very short distance from the transmitting station.

AMATEUR RADIO PRACTICES

Station Operation and set up

Modern HF transceivers have features that make operating a breeze, but to use them properly, you have to know when to use them and how to use them. The notch filter is a good example. The purpose of the "notch filter" found on many HF transceivers is **to reduce interference from carriers in the receiver passband**. (G4A01)

Another feature that helps reduce interference from nearby stations is the IF shift control. It shifts the passband of the IF filter to the right or left of the center frequency. One use for the IF shift control on a receiver is **to avoid interference from stations very close to the receive frequency**. (G4A11)

One type of interference is called overload. This occurs when a strong incoming signal is close to the frequency that you're monitoring. To help prevent this type of interference, many transceivers have an attenuator, which you can switch in to reduce the signal level reaching the RF amplifiers. One reason to use the attenuator function that is present on many HF transceivers is **to reduce signal overload due to strong incoming signals**. (G4A13)

Modern transceivers also have features that make operating CW more convenient and effective. The purpose of an electronic keyer, for example, is **automatic generation of strings of dots and dashes for**

CW operation. (G4A10) One advantage of selecting the opposite or "reverse" sideband when receiving CW signals on a typical HF transceiver is that **it may be possible to reduce or eliminate interference from other signals**. (G4A02)

Many modern transceivers also have the ability to receive on one frequency and send on another. This is called "split mode." Operating a transceiver in "split" mode means that **the transceiver is set to different transmit and receive frequencies**. (G4A03) Some high-end transceivers also have two or more variable frequency oscillators, or VFOs. A common use for the dual VFO feature on a transceiver is **to permit monitoring two different frequencies**. (G4A12) These features are especially useful when operating DX, that is when contacting foreign stations.

It has become quite popular to operate "digital modes" by supplying an audio signal, usually from a computer sound card, to a transceiver set to transmit single sideband. This type of operation is called audio frequency shift keying, or AFSK, because the audio tones will shift the transmitted frequency.

In order to send a clean signal, you must set the mic gain and automatic level control (ALC) properly. If the ALC system is not set properly when transmitting AFSK signals with the radio using single sideband mode, **improper action of ALC distorts the signal and can cause spurious emissions**. (G4A14)

One problem that may occur when tranmitting AFSK signals is that the audio cable connecting the computer to the radio may pick up stray RF. **All of these choices are correct**, i.e. can be a symptom of transmitted RF being picked up by an audio cable carrying AFSK data signals between a computer and a transceiver (G4A15):
- The VOX circuit does not un-key the transmitter
- The transmitter signal is distorted
- Frequent connection timeouts

Many amateurs buy linear amplifiers to make their signals stronger. Knowing how to use these devices is important so that you transmit clean signals and avoid interfering with other amateur radio stations.

The correct adjustment for the load or coupling control of a vacuum tube RF power amplifier is to adjust for **maximum power output without exceeding maximum allowable plate current**. (G4A08) **A pronounced dip** on the plate current meter of a vacuum tube RF power amplifier indicates correct adjustment of the plate tuning control. (G4A04)

The purpose of using Automatic Level Control (ALC) with a RF power amplifier is **to reduce distortion due to excessive drive**. (G4A05) **Excessive drive power** is a condition that can lead to permanent damage when using a solid-state RF power amplifier. (G4A07)

A time delay is sometimes included in a transmitter keying circuit **to allow time for transmit-receive changeover operations to complete properly before RF output is allowed**. (G4A09)

Antenna tuners, also known as antenna couplers, are also common accessories in an amateur radio station. An **antenna coupler or antenna tuner** is often used to enable matching the transmitter output to an impedance other than 50 ohms. (G4A06)

Test and monitoring equipment, two-tone test

When you set up your amateur radio station, sometimes called your "shack," you'll not only want to acquire radios, but also some test equipment. The most basic piece of test equipment is the voltmeter.

Voltmeters may be either analog or digital, but most amateurs now choose digital meters because they are cheaper and more accurate than analog meters. An advantage of a digital voltmeter as compared to an analog voltmeter is that it has **better precision for most uses**. (G4B06) Another advantage is high input impedance. High input impedance is desirable for a voltmeter because **it decreases the loading on circuits being measured**. (G4B05)

The use of an analog meter might, however, be preferred over a digital meter in some applications. **When adjusting tuned circuits,** the use of an instrument with analog readout may be preferred over an instrument with a numerical digital readout. (G4B14) The reason for this is that with an analog meter you can more easily see how a circuit's output changes as you tune it.

An oscilloscope is another handy piece of test equipment to have in your shack. With an oscilloscope, you can view signal waveforms. **An oscilloscope** is an item of test equipment that contains horizontal and vertical channel amplifiers. (G4B01) One advantage of an oscilloscope versus a digital voltmeter is that **complex waveforms can be measured**. (G4B02)

An oscilloscope is the best instrument to use when checking the keying waveform of a CW transmitter. (G4B03) **The attenuated RF output of the transmitter** is the signal source that is connected to the vertical input of an oscilloscope when checking the RF envelope pattern of a transmitted signal. (G4B04)

Antenna analyzers are instruments that can measure a number of different parameters associated with antennas, such as SWR. The **antenna and feed line** must be connected to an antenna analyzer when it is being used for SWR measurements. (G4B11) A use for an antenna analyzer, other than measuring the SWR of an antenna system, is **determining the impedance of an unknown or unmarked coaxial cable**. (G4B13)

Strong signals from nearby transmitters can affect the accuracy of measurements when making measurements on an antenna system with an antenna analyzer. (G4B12) This is because the antenna being analyzed will pick up RF energy from the nearby transmitters, and this energy will be read as excessive reflected power.

Standing wave ratio can also be determined with a directional wattmeter. (G4B10) To measure the SWR with a direction wattmeter, you first measure the power in one direction, then in the opposite direction, and finally calculate the SWR.

Another instrument often used for making antenna measurements is the field strength meter. **The radiation pattern of an antenna** can be determined with a field strength meter. (G4B09) **A field-strength meter** may also be used to monitor relative RF output when making antenna and transmitter adjustments. (G4B08)

One test that is often run on a SSB transmitter is the two-tone test. A two-tone test analyzes the **linearity** of a transmitter. (G4B15) **Two non-harmonically related audio signals** are used to conduct a two-tone test. (G4B07)

Interference with consumer electronics, grounding, DSP

At some point or another, your amateur radio station will interfere with a radio, television set, or telephone. Sometimes this may be your fault, other times it may be the fault of the device. In either case, you should do everything you can to eliminate this interference.

Public-address (PA) systems and telephones are often targets of interference. **Distorted speech** is heard from an audio device or telephone if there is interference from a nearby single-sideband phone transmitter. (G4C03) **On-and-off humming or clicking** is the effect that a nearby CW transmitter may have on an audio device or telephone system. (G4C04)

Fortunately, there are many things you can do to reduce or eliminate the interference. For example, a **bypass capacitor** might be useful in reducing RF interference to audio-frequency devices. (G4C01) You can also use ferrite chokes. **Placing a ferrite choke around the cable** would reduce RF interference caused by common-mode current on an audio cable. (G4C08)

Proper grounding is also important. One good way to avoid unwanted effects of stray RF energy in an amateur station is to **connect all equipment grounds together**. (G4C07)

Rather than connecting them in a daisy-chain fashion, you should connect them all to a single point. **Connect all ground conductors to a single point** to avoid a ground loop. (G4C09) If **you receive reports of "hum" on your station's transmitted signal,** this could be a symptom of a ground loop somewhere in your station. (G4C10)

A long ground wire will act more like an antenna at high frequencies than it will at DC or low frequencies. As a result, it is important to keep ground connections as short as possible to prevent high-impedance or resonant ground connections.

One effect that can be caused by a resonant ground connection is **high RF voltages on the enclosures of station equipment**. (G4C06) If you receive an RF burn when touching your equipment while transmitting on an HF band, assuming the equipment is connected to a ground rod, **the ground wire has high impedance on that frequency**. (G4C05)

A common complaint of amateur radio operators is electrical noise that seems to be on every band. This may be caused by arcing in a power line transformer or at some other connection. **Arcing at a poor electrical connection** could be a cause of interference covering a wide range of frequencies. (G4C02)

One thing you might do to reduce or eliminate this interference is use a digital signal processor, or DSP. One function of a digital signal processor in an amateur station is **to remove noise from received signals**. (G4C11)

You can also use DSPs to eliminate interference from amateur radio signals that are close to the frequency you are operating on. **A Digital Signal Processor (DSP) filter** can perform automatic notching of interfering carriers. (G4C13)

Many modern amateur radio transceivers have built-in DSPs that operate at IF frequencies, or you can purchase speakers designed for communications use that have audio DSPs. An advantage of a receiver DSP IF filter as compared to an analog filter is that **a wide range of filter bandwidths and shapes can be created**. (G4C12)

Speech processors, S meters, sideband operation near band edges

Speech processors can be very useful when operating SSB. The purpose of a speech processor as used in a modern transceiver is to **increase the intelligibility of transmitted phone signals during poor conditions**. (G4D01) **It increases average power** of a transmitted single sideband phone signal. (G4D02)

Of course, you must adjust it properly to gain these benefits. **All of these choices are correct** when considering the effects of an incorrectly adjusted speech processor (G4D03):

- Distorted speech
- Splatter
- Excessive background pickup

Most commercial receivers have an S meter. An S meter is found **in a receiver**. (G4D06) An S meter measures **received signal strength**. (G4D04)

The S meter uses a logarithmic scale, with an increase of one S unit being equivalent to a gain of 6 dB. So, to change the S meter reading on a distant receiver from S8 to S9, you would have to raise the power output of your transmitter **approximately 4 times**. (G4D07) Assuming a properly calibrated S meter, an S meter reading of 20 dB over S-9 **is 100 times stronger** compared to an S-9 signal. (G4D05)

When operating near band edges, it's important to know the frequency range that your signal will actually occupy so that your transmissions stay within the amateur band. When operating in LSB mode, your signal actually occupies a 3 kHz space below the displayed carrier frequency. When the displayed carrier frequency is set to 7.178 MHz, a 3 kHz LSB signal occupies **7.175 to 7.178 MHz**. (G4D08)

When operating in the 40 meter General Class phone segment when using 3 kHz wide LSB, your displayed carrier frequency should be at least **3 kHz above the edge of the segment**. (G4D10) That is to say that you should not set your radio to a frequency less than 7.178 MHz so that your transmitted signal stays within the General Class portion of the band.

Similarly, when operating USB, your signal occupies a space starting at the displayed carrier frequency and extending up 3 kHz. With the displayed carrier frequency set to 14.347 MHz, a 3 kHz USB signal occupies **14.347 to 14.350 MHz**. (G4D09) When operating in the 20 meter General Class band, your displayed carrier frequency should be no closer than **3 kHz below the edge of the band** when using 3 kHz wide USB. (G4D11)

HF mobile radio installations; emergency and battery powered operation

Operating mobile, that is from a car or boat, is an activity enjoyed by many radio amateurs. Amateurs can do almost everything from a mobile station that they can do at a fixed station.

When setting up a mobile station, the first thing to consider is how you are going to supply power to your radios. A direct, fused power connection **to the battery using heavy gauge wire** would be the best for a 100-watt HF mobile installation. (G4E03) It is best NOT to draw the DC power for a 100-watt HF transceiver from an automobile's auxiliary power socket because **the socket's wiring may be inadequate for the current being drawn by the transceiver**. (G4E04)

One common complaint when operating mobile is that the vehicle's electronics generate noise that interferes with sensitive HF receivers. **All of these choices are correct** when talking about what may cause interference to be heard in the receiver of an HF mobile installation in a recent model vehicle (G4E07):

- The battery charging system
- The fuel delivery system
- The vehicle control computer

As you can imagine, mobile antennas for HF operation are always a compromise because they are so short with respect to wavelength. **The antenna system** is the one thing that most limits the effectiveness of an HF mobile transceiver operating in the 75 meter band. (G4E05) One disadvantage of using a shortened mobile antenna as opposed to a full size antenna is that **operating bandwidth may be very limited**. (G4E06)

One way to make a physically short antenna resonate on HF is to use a capacitance hat. The purpose of a capacitance hat on a mobile antenna is **a device to electrically lengthen a physically short antenna**. (G4E01)

Finally, because mobile antennas are so short, the voltage at the tip of the antenna may be very high. The purpose of a corona ball on a HF mobile antenna is **to reduce high voltage discharge from the tip of the antenna**. (G4E02)

Some amateurs use solar cells to provide emergency power. The process by which sunlight is changed directly into electricity is called **photovoltaic conversion**. (G4E08) The approximate open-circuit voltage from a modern, well-illuminated photovoltaic cell is **0.5 VDC**. (G4E09)

Some amateurs use photovoltaic cells to recharge storage batteries. These systems often have a series diode connected between a solar panel and the battery. The reason a series diode is connected between a solar panel and a storage battery being charged by the panel is that **the diode prevents self discharge of the battery though the panel during times of low or no illumination**. (G4E10)

Some amateurs even use wind power as a power source. One disadvantage of using wind as the primary source of power for an emergency station is that **a large energy storage system is needed to supply power when the wind is not blowing**. (G4E11)

OPERATING PROCEDURES

Phone operating procedures, USB/LSB utilization conventions, procedural signals, breaking into a QSO in progress, VOX operation

Single sideband is the mode of voice communication most commonly used on the high frequency amateur bands. (G2A05) When using single sideband (SSB) voice mode, **only one sideband is transmitted; the other sideband and carrier are suppressed.** (G2A07) Because only one sideband is transmitted, **less bandwidth used and greater power efficiency** is an advantage when using single sideband as compared to other analog voice modes on the HF amateur bands. (G2A06)

Upper sideband is the sideband most commonly used for voice communications on frequencies of 14 MHz or higher. (G2A01) Accordingly, **upper sideband** is most commonly used for voice communications on the 17 and 12 meter bands (G2A04) and **upper sideband** is most commonly used for SSB voice communications in the VHF and UHF bands. (G2A03)

Lower sideband is the mode most commonly used for voice communications on the 160, 75, and 40 meter bands. (G2A02) The reason most amateur stations use lower sideband on the 160, 75 and 40 meter bands is that **current amateur practice is to use lower**

sideband on these frequency bands. (G2A09)

To establish a contact on the HF bands, you can call CQ or reply to a CQ. Sometimes you'll hear stations calling "CQ DX" instead of simply CQ. The expression "CQ DX" usually indicates **the caller is looking for any station outside their own country**. (G2A11)

Another way to establish a contact is to break into a conversation that's already in progress. The recommended way to break into a conversation when using phone is to **say your call sign during a break between transmissions from the other stations.** (G2A08)

When operating SSB, many amateurs like to use the VOX, or voice-operated control, instead of push-to-talk, or PTT, operation . Why would amateurs use voice VOX operation versus PTT operation? **VOX allows "hands free" operation**. (G2A10)

Operating courtesy; band plans; emergencies, including drills and emergency communications

Whenever you're operating, courtesy should always be a consideration when selecting a frequency. **Except during FCC declared emergencies, no one has priority access to frequencies, common courtesy should be a guide**. (G2B01)

One of the ways to be courteous is to select your operating frequency so that you do not interfere with other stations operating on nearby frequencies. The first thing you should do is make sure that you choose an appropriate frequency for the mode you are going to use. To comply with good amateur practice when choosing a frequency on which to initiate a call, **follow the voluntary band plan for the operating mode you intend to use**. (G2B07)

Some band plans may denote a frequency or small band of frequencies as the DX window for that band. The "DX window" in a voluntary band plan is **a portion of the band that should not be used for contacts between stations within the 48 contiguous United States**. (G2B08)

Next, avoid interference with other stations by ensuring that the frequency you want to use is not already in use. Remember that on many HF bands you may only be able to hear one station and not the other, so before calling, ask if the frequency is in use. A practical way to avoid harmful interference on an apparently clear frequency before calling CQ on CW or phone is to **send "QRL?" on CW, followed by your call sign; or, if using phone, ask if the frequency is in use, followed by your call sign**. (G2B06)

Another thing you should do is to make sure that the frequency you wish to use is not too close to other stations. The customary minimum frequency separation between SSB signals under normal conditions is **approximately 3 kHz**. (G2B05) When selecting a CW transmitting frequency, **150 to 500 Hz** is the minimum separation that should be used to minimize interference to stations on adjacent frequencies. (G2B04)

Band conditions can change during the course of a contact. If propagation changes during your contact and you notice increasing interference from other activity on the same frequency, **as a common courtesy, move your contact to another frequency**. (G2B03)

It's also important to know what to do if you hear a station in an emergency situation. While the rules are very strict regarding normal operation of an amateur radio station, an amateur station is allowed to use any means at its disposal to assist another station in distress **at any time during an actual emergency**. (G2B12) This means using frequencies outside of the amateur bands and using high power should the situation require it. If you find yourself in an emergency situation, you should send a distress call on **whichever frequency has the best chance of communicating the distress message**. (G2B11)

The first thing you should do if you are communicating with another amateur station and hear a station in distress break in is to **acknowledge the station in distress and determine what assistance may be needed**. (G2B02) Helping that station in distress should become your first priority.

In certain emergencies, government officials might activate the Radio Amateur Civil Emergency Service (RACES). **Only a person holding an FCC issued amateur operator license** may be the control operator of an amateur station transmitting in RACES to assist relief operations during a disaster. (G2B09) If the situation is really dire, more specifically **when the President's War Emergency Powers have been invoked,** the FCC may restrict normal frequency operations of amateur stations participating in RACES. (G2B10)

CW operating procedures and procedural signals, Q signals and common abbreviations, full break in

Just like phone operation, you can establish contact with another station using CW by either calling CQ and waiting for others to call you or by listening for other stations calling CQ and answering them. If you decide to call CQ, you should send "QRL?" before transmitting. The Q-signal "QRL" means **"Are you busy?", or "Is this frequency in use?"** (G2C04)

If you decide to answer a CQ, try to send at the same speed as the station calling CQ. The best speed to use answering a CQ in Morse Code is **the speed at which the CQ was sent.** (G2C05) Accordingly, if someone answers your CQ at a slower speed at which you sent it, slow down to match the speed of the station calling you.

When answering a CQ, you should set the frequency of your transceiver so that it matches the frequency of the sending station. We call this process "zero beating." The term "zero beat" in CW operation means **matching your transmit frequency to the frequency of a received signal**. (G2C06)

After establishing contact, it's customary to send the other station a signal report. This report consists of three numbers that correspond to the readability, strength, and tone of the signal. Hams sometimes call this the RST report. When sending CW, a "C" added to the RST report means a **chirpy or unstable signal**. (G2C07)

When operating CW, it's a good idea to use Q signals to shorten the number of characters that you must send and which the receiving station needs to receive. Q signals are three letter combinations that begin with the letter "Q."

There are many different Q signals, but you only need to know four of them to pass the test. The Q signal "QSL" means "**I acknowledge receipt.**" (G2C09) The Q signal "QRN" means "**I am troubled by static.**" (G2C10) The Q signal "QRV" means "**I am ready to receive messages**." (G2C11)If a CW station sends "QRS", **send slower**. (G2C02)

CW operators also use what are called prosigns. **AR** is the prosign sent to indicate the end of a formal message when using CW. (G2C08) When a CW operator sends "KN" at the end of a transmission, it means the operator is **listening only for a specific station or stations**. (G2C03)

When operating CW, some hams simply mute their receivers during a transmission. The problem with this approach is that the other operator cannot break in to make a comment. Another problem is that you can't hear if another station is interfering with your transmission. To get around these problems, hams use "break-in mode." When using full break-in telegraphy (QSK), **transmitting stations can receive between code characters and elements**. (G2C01)

Amateur Auxiliary, minimizing interference, HF operations

This section is really just a bunch of miscellaneous questions relating to operating an amateur radio station.

The FCC's Amateur Auxiliary is a group of **amateur volunteers who are formally enlisted to monitor the airwaves for rules violations**. (G2D01) Hams that are ARRL Official Observers are part of the Amateur Auxiliary. To become a member, you must take and pass a test.

The objectives of the Amateur Auxiliary are **to encourage self regulation and compliance with the rules by radio amateurs**. (G2D02) **Direction finding used to locate stations violating FCC Rules** is one of the skills learned during "hidden transmitter hunts" that are of help to the Amateur Auxiliary. (G2D03)

It is permissible to communicate with amateur stations in countries outside the areas administered by the Federal Communications Commission **when the contact is with amateurs in any country except those whose administrations have notified the ITU that they object to such communications**. (G2D05) That's a long way of telling you not to talk to hams where the governments don't want their hams talking to foreigners. In practice, if a country has in place such a restriction, they probably don't have many licensed hams.

There are three questions on using directional antennas. **A directional antenna** would be the best HF antenna to use for minimizing interference. (G2D11) The reason for this is that you can point the antenna away from a device you are interfering with.

Most communications take place on the "short path," that is the most direct path between two stations. At times, however, propagation may favor the long path. A directional antenna is pointed **180 degrees from its short-path heading** when making a "long-path" contact with another station. (G2D06)

To figure where to point a directional antenna you'd use an

azimuthal projection map. An azimuthal projection map is **a map that shows true bearings and distances from a particular location**. (G2D04)

When I got my license, you had to log every transmission you made, even if you were just calling CQ. Nowadays, there's no need to keep a log, except in some very specific instances, but even though it's not required, it still might be a good idea. One reason many amateurs keep a log even though the FCC doesn't require it is **to help with a reply if the FCC requests information**. (G2D08)

All of these choices are correct when talking about what information is traditionally contained in a station log (G2D09):

- Date and time of contact
- Band and/or frequency of the contact
- Call sign of station contacted and the signal report given

One of the situations that does require that you log certain information is when you are operating in the 60-meter band. When operating in the 60 meter band, the FCC rules require that, **if you are using other than a dipole antenna, you must keep a record of the gain of your antenna**. (G2D07)

QRP operation refers to **low power transmit operation**. (G2D10)

Digital operating: procedures, procedural signals, and common abbreviations

In recent years, operating what's known as the "digital modes" has become popular. They are known by this name because you can't operate them without a computer. RTTY, PSK31, JT65, and PACTOR are examples of digital modes.

When operating the digital modes, amateurs connect the audio output of their transceivers into the input of a computer sound card and the output of the sound card to the audio input of the transceiver. The computer does all the heavy lifting, decoding the tones on the input and generating the tones to be transmitted.

Amateurs operate these modes in specific sub-bands. **14.070 - 14.100 MHz** is the segment of the 20 meter band that is most often used for data transmissions. (G2E04) **Below the RTTY segment, near 14.070 MHz** is the segment of the 20 meter band where most PSK31 operations are commonly found. (G2E08) **3585 – 3600 kHz** is the segment of the 80 meter band most commonly used for digital transmissions. (G2E07)

Amateurs have actually been operating RTTY for a long time, but it has become a lot more popular in recent years because amateurs can now operate this mode using a computer and not a clunky, mechanical teletype machine. RTTY signals shift between two frequencies. That's why this mode is called frequency shift keying. One frequency denotes a "1," while another denotes a "0." The difference between those two frequencies is called the frequency shift. **170 Hz** is the most common frequency shift for RTTY emissions in the amateur HF bands. (G2E06)

Using a computer sound card to generate audio tones that shift the frequency of a transmitted signal is called audio frequency shift keying (AFSK). **LSB** is the mode normally used when sending an RTTY signal via AFSK with an SSB transmitter. (G2E01) **USB** is the standard sideband used to generate a JT65 or JT9 digital signal when using AFSK in any amateur band. (G2E05)

To tune and decode digital signals, amateur radio operators use computer programs that display signals in a portion of an amateur radio band in what is called a waterfall display. In a waterfall display, **frequency is horizontal, signal strength is intensity, time is vertical**. (G2E12) When an operator selects one of the signals being displayed, the computer will then begin to decode that signal.

By looking carefully at the signal displayed on the waterfall, you can tell certain things about the signal. For example, one or more vertical lines adjacent to a PSK31 signal indicates **overmodulation**. (G2E11)

Properly tuning a digital signal is important. **All of these choices are correct** when deciding what could be wrong if you cannot decode an RTTY or other FSK signal even though it is apparently tuned in properly: (G2E14)

- The mark and space frequencies may be reversed
- You may have selected the wrong baud rate
- You may be listening on the wrong sideband

Many digital modes, such as RTTY and PSK31, have no error correction. Occasionally, there are errors receiving these signals, but generally, that's not a problem, as operators can generally fill in the blanks.

If errors cannot be tolerated, you might want to consider using a mode called PACTOR. PACTOR uses the ARQ protocol for error correction, and a network of stations running PACTOR has been set up to relay data between radio stations and the Internet. This communication system, called **Winlink**, sometimes uses the Internet to transfer messages. (G2E13)

Stations that are connected to the Internet are called gateways. To establish contact with a digital messaging system gateway station, **transmit a connect message on the station's published frequency**. (G2E10) Be aware that there is no way to join a contact between two stations using the PACTOR protocol. **Joining an existing contact is not possible, PACTOR connections are limited to two stations.** (G2E09)

Like any amateur radio communications, you should make sure that a frequency is not in use before starting a PACTOR transmission. To do this, **put the modem or controller in a mode which allows monitoring communications without a connection** to determine if the channel is in use by other PACTOR stations. (G2E02)

Even though the PACTOR protocol includes error correction, it's still a good idea to choose a clear frequency so that there is no interference. **All of these choices are correct** when talking about symptoms that may result from other signals interfering with a PACTOR or WINMOR transmission (G2E03):

- Frequent retries or timeouts
- Long pauses in message transmission
- Failure to establish a connection between stations

ELECTRICAL AND RF SAFETY

RF safety principles, rules and guidelines, routine station evaluation

If you learn nothing else from this manual, I hope that you learn to be safe when setting up your station, building antennas, or operating a radio. It's unfortunate, but hams every year lose their lives in accidents that could have been prevented. In this chapter, we will cover RF safety and electrical safety.

By RF safety, we mean safe exposure to the RF energy generated by amateur radio transmitters. One way that RF energy can affect human body tissue is that **it heats body tissue**. (G0A01) In reference to RF radiation exposure, "time averaging" means **the total RF exposure averaged over a certain time**. (G0A04)When evaluating RF exposure, **a lower transmitter duty cycle permits greater short-term exposure levels**. (G0A07)

All of these choices are correct when talking about properties that are important in estimating whether an RF signal exceeds the maximum permissible exposure (MPE) (G0A02):
- Its duty cycle
- Its frequency
- Its power density

If you install an indoor transmitting antenna, **make sure that**

MPE limits are not exceeded in occupied areas. (G0A11)

How do you know if you're being exposed to higher levels of RF radiation than you should be? One way is to measure the RF field. **A calibrated field-strength meter with a calibrated antenna** can be used to accurately measure an RF field. (G0A09)

To ensure compliance with RF safety regulations when transmitter power exceeds levels specified in part 97.13, an amateur operator must **perform a routine RF exposure evaluation**. (G0A08) **All of these choices are correct** for ways that you can determine that your station complies with FCC RF exposure regulations (G0A03):

- By calculation based on FCC OET Bulletin 65
- By calculation based on computer modeling
- By measurement of field strength using calibrated equipment

If an evaluation of your station shows RF energy radiated from your station exceeds permissible limits, you should **take action to prevent human exposure to the excessive RF fields**. (G0A05) If an evaluation shows that a neighbor might receive more than the allowable limit of RF exposure from the main lobe of a directional antenna, **take precautions to ensure that the antenna cannot be pointed in their direction**. (G0A10)

In addition to ensuring that you're not exposed to high-energy RF fields, there are some simple precautions that you should take when installing and tuning an antenna. **Turn off the transmitter and disconnect the feed line** whenever you make adjustments or repairs to an antenna. (G0A12) When installing a ground-mounted antenna, **it should be installed such that it is protected against unauthorized access**. (G0A06)

Following safe work practices when installing or working on an antenna tower is paramount. When climbing on a tower using a safety belt or harness, **confirm that the belt is rated for the weight of the climber and that it is within its allowable service life**. (G0B07)

Any person preparing to climb a tower that supports electrically powered devices should **make sure all circuits that supply power to**

the tower are locked out and tagged. (G0B08) Soldered joints should not be used with the wires that connect the base of a tower to a system of ground rods because **a soldered joint will likely be destroyed by the heat of a lightning strike**. (G0B09) A good practice for lightning protection grounds is that **they must be bonded together with all other grounds**. (G0B11)

Safety in the ham shack: electrical shock and treatment, safety grounding, fusing, interlocks, wiring, antenna and tower safety

When wiring a "shack," pay special attention to the currents that the circuit must supply and use the appropriate wire sizes and fuse sizes. **Electrical safety inside the ham shack** is covered by the National Electrical Code. (G0B14)

According to the code, **AWG number 12** is the minimum wire size that may be safely used for a circuit that draws up to 20 amperes of continuous current. (G0B02) **15 amperes** is the size of fuse or circuit breaker that would be appropriate to use with a circuit that uses AWG number 14 wiring. (G0B03)

For some devices, such as a linear amplifier, you may have to install a 240 VAC circuit. When doing so, remember **only the two wires carrying voltage** in a four-conductor connection should be attached to fuses or circuit breakers in a device operated from a 240-VAC single-phase source. (G0B01) **Current flowing from one or more of the voltage-carrying wires directly to ground** will cause a Ground Fault Circuit Interrupter (GFCI) to disconnect the 120 or 240 Volt AC line power to a device. (G0B05)

Another way to make the shack safer, is to properly ground your equipment. The metal enclosure of every item of station equipment must be grounded because **it ensures that hazardous voltages cannot appear on the chassis**. (G0B06)

Some equipment may have features that make it safer to use or maintain. For example, some power supplies have interlock switches. The purpose of a transmitter power supply interlock is **to ensure that dangerous voltages are removed if the cabinet is opened**. (G0B12)

Finally, consider that the lead in the solder commonly used in amateur radio equipment might constitute a hazard. One danger from lead-tin solder is that **lead can contaminate food if hands are not washed carefully after handling**. (G0B10)

In an emergency, you may want to power your station with batteries or an emergency power generator. Again, please do so safely.

When powering your house from an emergency generator, you must **disconnect the incoming utility power feed**. (G0B13)

Gasoline-powered generators may emit exhaust gases that could be hazardous. For an emergency generator installation, **the generator should be located in a well ventilated area**. (G0B15) **Danger of carbon monoxide poisoning** is a primary reason for not placing a gasoline-fueled generator inside an occupied area. (G0B04)

COMMISSION'S RULES

General Class control operator frequency privileges, primary and secondary allocations

As you know, on the 80m, 40m, 20m, and 15m bands, some frequencies are reserved for Advanced and Extra Class licensees. On the other bands, however, General Class licensees have exactly the same privileges as Advanced and Extra Class licensees. **160, 60, 30, 17, 12, and 10 meters** are the bands that a General Class license holder is granted all amateur frequency privileges. (G1A01)

On 80 meters, General Class operators may only operate SSB above 3.800 MHz, so **3900 kHz** is a frequency that is within the General Class portion of the 75 meter phone band. (G1A06) General Class licensees cannot operate CW below 3.525 MHz. That means that **3560 kHz** is a frequency that is within the General Class portion of the 80 meter band. (G1A08)

On 40m, General Class operators may only operate SSB from 7.175 to 7.300 MHz. That means **7.250 MHz** is a frequency in the General Class portion of the 40 meter band. (G1A05)

On 20m, General Class licensees can only operate phone above 14.225 MHz. So, **14305 kHz** is a frequency within the General Class portion of the 20 meter phone band. (G1A07)

On 15m, the General Class portion of the phone band is 21.275

to 21.450 MHz. That means **21300 kHz** is a frequency that is within the General Class portion of the 15 meter band. (G1A09)

Can you see a pattern emerging here? When General Class licensees are not permitted to use the entire voice portion of a particular band, **the upper frequency end** is the portion of the voice segment that is generally available to them. (G1A11)

On 10m, Advanced and Extra Class licensees have no special privileges. So, **all these answers are correct** when talking about frequencies available to a control operator holding a General Class license (G1A10):

- 28.020 MHz
- 28.350 MHz
- 28.550 MHz

30 meters has some restrictions that you won't find on other bands. For example, **30 meters** is the band on which phone operation is prohibited (G1A02), and **30 meters** is the band on which image transmission is prohibited. (G1A03)

60 meters is also kind of an oddball band. **60 meters** is the amateur band that is restricted to communication on only specific channels, rather than frequency ranges. (G1A04) This is the only band where amateur radio operators are restricted to specific frequencies.

On both the 30-meter band and the 60-meter band, amateur radio is a secondary user. When the FCC rules designate the Amateur Service as a secondary user on a band, **amateur stations are allowed to use the band only if they do not cause harmful interference to primary users**. (G1A12) When operating on either the 30 or 60 meter bands, and a station in the primary service interferes with your contact, the appropriate action is to **move to a clear frequency or stop transmitting**. (G1A13)

Amateurs in different ITU regions have different frequency privileges. In **Region 2**, the region in which most of the U.S. is located, operation in the 7.175 to 7.300 MHz band is permitted for a control operator holding an FCC issued General Class license.

(G1A14) In ITU Region 1, the region that encompasses most of Europe, amateurs are only allowed to operate up to 7.200 MHz.

Antenna structure limitations, good engineering and good amateur practice, beacon operation, restricted operation, retransmitting radio signals

The FCC also has some rules and regulations regarding antennas and operating practices. For example, **200 feet** is the maximum height above ground to which an antenna structure may be erected without requiring notification to the FAA and registration with the FCC, provided it is not at or near a public use airport. (G1B01)

Normally, amateur radio transmissions are only allowed when a two-way communication is taking place. The FCC does allow amateurs to set up beacon stations, though. Beacon stations continually transmit a low-power signal, often containing station location information. **Observation of propagation and reception** is a purpose of a beacon station as identified in the FCC Rules. (G1B03) One of the conditions with which beacon stations must comply is that **there must be no more than one beacon signal transmitting in the same band from the same station location**. (G1B02) The power limit for beacon stations is **100 watts PEP output**. (G1B10)

Because the rules state that amateur radio stations may not be used for any commercial purposes, you must be careful to whom you pass messages, and you should know what they plan to do with that information before you do. Before amateur stations may provide communications to broadcasters for dissemination to the public, **the communications must directly relate to the immediate safety of human life or protection of property and there must be no other means of communication reasonably available before or at the time of the event**. (G1B04)

Similarly, only **when other amateurs are being notified of the sale of apparatus normally used in an amateur station and such activity is not done on a regular basis** may an amateur station transmit communications in which the licensee or control operator has a pecuniary (monetary) interest. (G1B09) Basically what this means is that you can tell other hams that you have some personal gear to sell on your club's 2m net, but you can't set up a business selling used gear and use that net to advertise what you have available.

Music is not allowed on amateur radio, except for one specific circumstance. You can't even sing happy birthday to a friend over amateur radio! Only **when it is an incidental part of a manned space craft retransmission** may music be transmitted by an amateur station. (G1B05)

Likewise, secret codes are not allowed to be transmitted by an amateur radio station, except in one particular circumstance. An amateur station is permitted to transmit secret codes **to control a space station**. (G1B06) FCC rules define a space station as "an amateur station located more than 50 km above the Earth's surface." Similarly, a restriction on the use of abbreviations or procedural signals in the Amateur Service is that **they may be used if they do not obscure the meaning of a message**. (G1B07)

As long as the frequency is within the frequency sub-bands allotted to General Class operators, a General Class operator may choose to transmit on that frequency. However, when choosing a transmitting frequency, **all of these choices are correct** if you want to comply with good amateur practice (G1B08):

- Review FCC Part 97 Rules regarding permitted frequencies and emissions.
- Follow generally accepted band plans agreed to by the Amateur Radio community.
- Before transmitting, listen to avoid interfering with ongoing communication

Follow these guidelines and you'll avoid interfering with other stations.

The FCC rules do spell out certain technical requirements for your amateur radio station. They don't, however, cover every situation. In these cases, the FCC requires an amateur station to be operated **in conformance with good engineering and good amateur practice** in all respects not specifically covered by the Part 97 rules. (G1B11) **The FCC** determines "good engineering and good amateur practice" as applied to the operation of an amateur station in all respects not covered by the Part 97 rules. (G1B12)

Transmitter power regulations, data emission standards

In general, 1500 W PEP is the maximum output power an amateur may use when transmitting. **1500 watts PEP output** is, therefore, a limitation on transmitter power on the 28 MHz band for a General Class control operator (G1C05), **1500 watts PEP output** is a limitation on transmitter power on the 1.8 MHz band (G1C06), and **1500 watts PEP output** is the maximum transmitting power an amateur station may use on the 12 meter band. (G1C02)

In addition to this absolute power limit, the rules state that amateurs should use only enough power required to carry out a specific communication. Other than the 1500 watt PEP limit, the other limitation that applies to transmitter power on every amateur band is that **only the minimum power necessary to carry out the desired communications should be used**. (G1C04)

On 30 meters, though, the output power limit is lower because we are secondary users of the band. **200 watts PEP output** is the maximum transmitting power an amateur station may use on 10.140 MHz. (G1C01)

There are also some rules about bandwidth. For example, **2.8 kHz** is the maximum bandwidth permitted by FCC rules for Amateur Radio stations when transmitting on USB frequencies in the 60 meter band. (G1C03)

Similarly, there are rules related to data rates for the digital modes. I say similarly because the higher the data rate, the more bandwidth a signal will occupy.

300 baud is the maximum symbol rate permitted for RTTY or data emission transmitted at frequencies below 28 MHz. (G1C08) **300 baud** is also the maximum symbol rate permitted for RTTY or data emission transmission on the 20 meter band. (G1C07)

On 10 meters and above, there is more spectrum available, so higher data rates are allowed. **1200 baud** is the maximum symbol rate permitted for RTTY or data emission transmissions on the 10 meter band. (G1C10) **19.6 kilobaud** is the maximum symbol rate permitted for RTTY or data emission transmissions on the 2 meter band. (G1C11) **56 kilobaud** is the maximum symbol rate permitted for RTTY or data emission transmitted on the 1.25 meter and 70 centimeter bands. (G1C09)

Volunteer Examiners and Volunteer Examiner Coordinators, temporary identification

Back in the old days (when I got my amateur radio license), you had to visit an FCC office to take the General Class license test. Now, a corps of Volunteer Examiners (VEs) administer the tests and transmit the test results to the FCC.

Volunteer Examiners are accredited by **a Volunteer Examiner Coordinator**. (G1D07) For a non-U.S. citizen to be an accredited Volunteer Examiner, **the person must hold a U.S. Amateur Radio license of General Class or above**. (G1D08) There is an age limit, too. **18 years** is the minimum age that one must be to qualify as an accredited Volunteer Examiner. (G1D10)

When you are an accredited VE holding a General Class operator license, you may only administer the **Technician** examination. (G1D02) **An FCC General Class or higher license and VEC accreditation** is sufficient for you to be an administering VE for a Technician Class operator license examination. (G1D05)

There are a bunch of rules that govern how Volunteer Examiners and Volunteer Examiner Coordinators are to operate. A requirement for administering a Technician Class operator examination is that **at least three VEC accredited General Class or higher VEs must be present**. (G1D04) Having three Volunteer Examiners present helps ensure that the tests are administered properly and fairly.

When an applicant passes the General Class license examination, he or she is issued a Certificate of Successful Completion of Examination. A Certificate of Successful Completion of Examination (CSCE) is valid for exam element credit for **365 days**. (G1D09)

You may operate **on any General or Technician Class band segment** if you are a Technician Class operator and have a CSCE for

General Class privileges. (G1D03)

Whenever you operate using General Class frequency privileges, you must add the special identifier "AG" after your call sign if you are a Technician Class licensee and have a CSCE for General Class operator privileges, but the FCC has not yet posted your upgrade on its Web site. (G1D06)

Recently, the FCC changed the rules as to what happens when an amateur radio license expires. It used to be that if your license expired, and you were beyond the renewal grace period, you had to re-take the tests. If you wanted a Technician Class license, you had to take the Technician Class written exam (Element 2). If you wanted a General Class exam, you had to take the Technician and General Class exams (Elements 2 and 3), and if you wanted an Amateur Extra Class license, you had to take all three written exams (Elements 2, 3, and 4).

Now, however, **any person who can demonstrate that they once held an FCC issued General, Advanced, or Amateur Extra class license that was not revoked by the FCC** may receive credit for the elements represented by an expired amateur radio license. (G1D01) This does not include the Technician Class written exam (Element 2), though. If a person has an expired FCC issued amateur radio license of General Class or higher, **the applicant must pass the current element 2 exam** before they can receive a new license. (G1D11)

When they do pass the Element 2 exam, the FCC will issue an Amateur Extra Class license to applicants that previously held an Amateur Extra Class license and a General Class license to applicants that previously held either a General Class or an Advanced Class license. If this makes no sense to you, you're not alone. It makes no sense to me, either.

Control categories; repeater regulations; harmful interference; third party rules; ITU regions; automatically controlled digital station

The FCC rules specify when and how an amateur radio station can transmit messages for persons who are not licensed radio amateurs. These messages are called "third party traffic." For a non-licensed person to communicate with a foreign Amateur Radio station from a US amateur station at which a licensed control operator is present, **the foreign amateur station must be in a country with which the United States has a third party agreement**. (G1E08) Third party traffic is prohibited with **every foreign country, unless there is a third party agreement in effect with that country**, except for messages directly involving emergencies or disaster relief communications. (G1E07) **Only messages relating to Amateur Radio or remarks of a personal character, or messages relating to emergencies or disaster relief** may be transmitted by an amateur station for a third party in another country. (G1E05)

While most third-party traffic rules cover communication with amateurs in other countries, there is an important rule concerning third-party traffic within the U.S. In particular, amateurs may not pass third-party traffic for persons whose licenses have been revoked. **If the third party's amateur license had ever been revoked and not reinstated,** it would disqualify that third party from participating in stating a message over an amateur station. (G1E01)

The 10m band is the only HF band in which amateur radio operators can operate repeater stations. A 10 meter repeater may retransmit the 2 meter signal from a station having a Technician Class control operator, **only if the 10 meter repeater control operator holds at least a General Class license.** (G1E02)

All of these choices are conditions that require an Amateur Radio station to take specific steps to avoid harmful interference to other users or facilities (G1E04):

- When operating within one mile of an FCC Monitoring Station

- When using a band where the Amateur Service is secondary
- When a station is transmitting spread-spectrum emissions

In the event of interference between a coordinated repeater and an uncoordinated repeater, **the licensee of the uncoordinated repeater has primary responsibility to resolve the interference**. (G1E06)

In recent years, the FCC has changed part 97 to clarify how and where automatically controlled digital stations may operate in order to prevent harmful interference to other stations. **Automatically controlled digital station** is the FCC term for an unattended digital station that transfers messages to and from the Internet. (G1E11) Automatically controlled stations transmitting RTTY or data emissions may communicate with other automatically controlled digital stations **anywhere in the 1.25-meter or shorter wavelength bands, and in specified segments of the 80-meter through 2-meter bands**. (G1E13) **The station initiating the contact must be under local or remote control** to conduct communications with a digital station operating under automatic control outside the automatic control band segments. (G1E03) **Under no circumstances** are messages that are sent via digital modes exempt from Part 97 third party rules that apply to other modes of communication. (G1E12)

English is the language you must use when identifying your station if you are using a language other than English in making a contact using phone emission. (G1E09)

ABOUT THE AUTHOR

I have been a ham radio operator since 1971 and a radio enthusiast as long as I can remember. In addition to being an active CW operator on the HF bands:

- I blog about amateur radio at KB6NU.Com, one of the leading amateur radio blogs on the Internet.
- In addition to this General Class study guide, I have written study guides for the Technician Class and Extra Class exams. You can find the No-Nonsense Technician Class License Study Guide and the No-Nonsense Extra Class License Study Guide in PDF, Nook (ePub) and Kindle (Mobipocket) formats on my website at http://www.kb6nu.com/study-guides/. The Technician Class study guide is also available in paperback. The Kindle versions are available on Amazon, while the Nook versions are available on Barnes&Noble.
- I am the author of *21 Things to Do With your Amateur Radio License*, a book for those who have been recently licensed or just getting back into the hobby, and *The CW Geek's Guide to Having Fun with Morse Code*. A book for those who want to learn and use this great mode. These books are also available on my website (http://www.kb6nu.com/shop/) and Amazon and Barnes&Noble.
- I send out a monthly column to approximately 350 amateur radio clubs throughout North America for publication in their newsletters.
- I am the station manager for WA2HOM (http://www.

wa2hom.org), the amateur radio station at the Ann Arbor Hands-On Museum (http://www.aahom.org).

- I teach amateur radio classes around the state of Michigan.
- I serve as the ARRL Michigan Section Training Manager and conduct amateur radio leadership workshops for amateur radio club leaders in Michigan.

You can contact me by sending e-mail to cwgeek@kb6nu.com. If you have comments or question about any of the material in this book, I hope you will do so.